MODULAR BATTERY SWAP

MODULAR BATTERY SWAP

HOW WE CAN GET *EVERYBODY* DRIVING AN ELECTRIC CAR

Lou Shrinkle

Waterside Productions

Printed in the United States of America

First Printing, 2022

ISBN-13: 978-1-956503-77-7 print edition
ISBN-13: 978-1-956503-78-4 ebook edition

Waterside Productions
2055 Oxford Ave
Cardiff, CA 92007
www.waterside.com

Dedicated to my grandsons, Logan and Cody, and the planet they will inherit; and to my wife, Patty; my daughter, Susan; my son, Steven; and all the others who believed in this invention and helped me with this endeavor in so many ways.

CONTENTS

PREFACE

We're living in quite a fascinating time. For the first time in our history, we're now faced with a climate emergency, on top of a seemingly endless pandemic and some crazy political nonsense. The last few years leave little doubt that something is happening with the climate, and it ain't good. Unfortunately, like many others at first, I was a little taken in by the climate skeptics. I hadn't made up my mind yet either way, and there was this sort of satisfaction with going against what I considered conventional climate wisdom. But then I started thinking that maybe it would be a good idea to try to slow down our emissions. I was still a teeny bit of a climate skeptic, but I knew that any reduction in greenhouse gases would typically also lower pollution, so I was okay with any decarbonization efforts since I've always been in favor of clean air and water.

But as I began the journey of working in battery storage technology, my full conversion occurred. After attending many talks at the Scripps Institution of Oceanography in La Jolla, California, I found it pretty telling how confident the scientists were about the progression of our climate disruption. *There was no debate.* And these were not out-of-touch nerds or corrupt bought-and-paid-for researchers, as one narrative would have you believe. No, they were like my neighbors and friends, but much, much smarter about how climate works—and also some of the most genuine people you could ever meet. Their curiosity about *me* was how I harbored any skepticism at all, and this made me think much more critically about climate change.

I became a member of SanDiego350.org, a climate activist group that has a specific goal of achieving 350 parts per million (ppm) CO_2 in the atmosphere, and I believe a goal such as this is very important. My only questions now about climate disruption have to do with how we react and adapt to the inevitable changes, exactly what those changes will be, and what we'll be able to do about it. The jury is already back, however, on whether there's a negative impact occurring today.

Just recently, I came extremely close—*two separate times*—to losing my small mountain cabin in the Sierra Nevada mountains. The worst thing is that it *did* destroy the homes of many of my neighbors, and we all lost a very substantial portion of our beautiful southern Sierra Nevada old-growth and redwood forests. Tragic, but this is a story about an invention: about all the events that came together to form an idea; how that idea evolved in fits and starts into its final form; and about sticking with that idea in the face of doubt, criticism, and even possible failure. It's also about how discovery leads to many other open doors and how inventions can grow into something bigger than the original intention.

So often through our technological history, we've seen that this is the way that invention and innovation work. The inventor of the laser probably had no idea of the myriads of applications that device would yield. So as I looked at modular battery-swap technology, I saw many benefits and applications that weren't really the reasons that initially motivated the invention. In this book, I'll also discuss those additional revelations.

I really do hope that a lot of you will find these advantages compelling enough to make modular battery swap part of the global conversation on how we'll get *everyone* driving an electric car.

PART I
BACKGROUND AND HISTORY

CHAPTER 1

2008–2011: THE EV FORMATIVE YEARS

To be sure, by 2008 the stage was pretty well set that there would be an electric vehicle (EV) business of some sort. Plug-in hybrid electric vehicles (PHEVs) were looking very promising, and the visionaries at Tesla (they were, in my view, *hyper*-visionaries) were starting the full EV revolution. I was working in the hard-disk-drive business on heads, disks, and read/write channels, which had ostensibly been my lifelong career since 1979. I'd studied physics and had received my bachelor's degree from the University of California, San Diego (UCSD) in 1974.

The development of read/write data channels and the work with disk heads and media was a whole fascinating story in itself. We saw *unbelievable* changes from a drive with 600Mbytes of data and the size of a washing machine, evolving to a drive with 2Tbytes (3,000 times more!) and the size of a hockey puck. This proved to me that if anything is possible to build, and if there's enough money to be made, then it *will* get built. We must keep this in mind as I talk about battery cell improvements later in the book. After I had around 30 patents issued and developed more than two dozen products, the hard-drive industry sadly started to stop growing, and I found myself working on proverbial buggy whips.

Some of my compadres in the hard-drive business who'd moved to China during the proliferation of hard-drive component manufacturing there also saw the writing on the wall. Some were looking at the battery industry as the new high-growth technology sector. I don't believe that they had any idea how massive the battery business would become.

Mike Chang was one of my work colleagues, and a good friend in the hard-drive business. He landed a job as president at Amperex Technology Ltd.

(ATL), a battery-manufacturing company in China. They were growing fast by making large numbers of 18650 battery cells for mostly laptop computers, along with lithium polymer cells for iPhones and similar-size devices. ATL wanted to expand into the EV market, although as I recall, there was much consternation—that market was a big unknown and looked a bit risky and difficult.

Mike contacted me, as well as a few other engineers in the States who all came from the hard-drive industry. None of us specifically had a lot of experience with EV batteries, but that was true with all of us early on when it came to hard-disk drives. I suppose that Mike was counting on our proven problem-solving skills and cockiness to help kick-start battery-pack developments. At the time I was in my late 50s, so I saw this as a way to do something that was really needed in the world. I was convinced that EVs could help get us to utopian levels of air quality. I felt privileged to be getting involved with vehicle electrification and gladly accepted the task.

The early work at ATL was hard—all those trips to China and getting used to working with a whole different group of engineers and chemists with all their own idiosyncrasies. We were dealing with the fast-changing goals of a company having difficulty defining how to build their EV battery business. At that time, the CEO of ATL was TH Chen. He had also worked with a number of us in the hard-drive business. Robin Zeng was the COO and is now the head of Contemporary Amperex Technology Ltd. (CATL)—likely the largest battery company in the world. I remember that Robin continually hammered away on the safety issues with Li-Ion cells, showing us these large sand buckets throughout the facility where workers could throw a burning cell to let 'em burn out. At the cigarette-smoke-filled meetings, TH stressed that people should not really "own" the batteries in their EVs. There was simply too much risk of loss, and the costs were just too high. I never forgot that statement—TH seemed pretty convinced of it. I learned a lot of stuff that stuck with me from *all* those guys at ATL.

However, the one area of the battery system design that came up often in meetings and discussions was how to get the battery pack in an EV "refueled" in a time comparable to refilling a gasoline car. This would require a full-charge time of around 5 minutes or less. The chemists at ATL kept insisting that this charge rate, or anything even in the ballpark, wasn't possible with the typical battery chemistries. I did a simple calculation and also figured that the power level from the charger would be in excess of 1.5MW! Both of these constraints were obvious showstoppers, so I quickly started to think outside that impossible box. I realized that some kind of battery swap had the potential to be fast enough or faster, and I started to put together a scenario of simplifying the swap by dividing up the battery into smaller modules. The full pack just seemed too dang big to me to easily handle in the exchange.

But this was a bit of a distraction, and we continued on with the work at hand. Nobody at ATL was too interested in my modular idea. I'd suggested that we could use wireless technology to communicate and simplify the pack wiring. Again, all this fell on deaf ears, and we just returned to the design of more "conventional" battery packs. I'd gone through this scenario before in the hard-disk world—I'd hear of a need that would require some out-of-the-box thinking, and I'd take on that challenge doggedly until I either came up with a new invention or realized its futility. I didn't ever think of myself as an "inventor," but as I look back, I guess I was. Usually, nobody I reported to ever asked me to come up with these specific inventions, and I initiated the work behind 90% of the patents. When I could see the necessity with clarity, I would then then set out to devise the new invention simply because it didn't exist. This creative act was, for me, its own reward.

We built numerous large, fixed battery packs. As we built them, I began noticing how difficult the development and prototyping of these large packs were. It was hard to find technicians or engineers among us who wanted to work on these packs. There was the fear of electrocution, and more than that, there was a true gripping fear of a thing called arc-spray. This is where any metal contacting across the high-voltage bus will explode in a blinding flash, sending molten metal at high speeds everywhere. Screwdriver blades just disappear!

The takeaway here for me was that these large high-powered packs were just very difficult to manage in development. Such extreme care had to be taken in *every* action, and progress would get painfully slow. They're heavy to move, and it's difficult to access portions of the pack for debugging and testing. It was looking again as if a move toward using *independent* modules was a much better, safer way. I realized then that somehow they had to be independent and physically attach and connect in a very forgiving manner. My concept of how to quickly refuel an electric car was beginning to take shape.

CHAPTER 2

2011–2012: THE BIRTH OF AN IDEA AND A YEAR OF TUMULT

Then came 2011—and what a year that was! First, at the end of 2010, TH, still CEO of ATL, was on a flight to San Jose when he fell ill. Sadly, he passed away from a very bad case of flu after arriving. In February 2011, a young engineer, Billy Wahng, and I made a trip to Beijing and Beijing Automobile Works, where the fledgling Chinese auto industry was still working out its bugs. It was quite a scene in retrospect, with prototype cars dressed up with cheesy vinyl wrap decals all over the place.

We spent many days working in an extremely large, chaotic garage. It was brutally cold on that trip—well below 10°F constantly. The garage was barely above freezing, the air was horribly polluted, and there was a film of greenish-brownish grunge on car windows and other surfaces. I started getting really sick and decided to get home quickly (not the best return flight). When I got back, my doctor said that my lungs were badly compromised, and he put me on a regimen of Cipro. On top of that, my partner, Billy, got a blood clot in his legs on the flight and spent some serious time in the hospital with further complications. Later that year in July, I was diagnosed with small-cell lymphoma. I can't say for sure that there was a connection, but who knows?

I decided that my traveling days to China were over. In 2011, all of a sudden there was a big change at ATL. Robin, Mike, and some others broke off from the company (which is owned by TDK, a Japanese company) and formed

CATL. Robin, Mike, and some others became multibillionaires at CATL, and deservedly so—they were some of the most focused and hardworking individuals I've ever been around.

I continued to do some consulting in the battery field. Then one day while reading the newspaper, I came across an article reporting that the city of Oceanside, California, was installing a significant number of Level 2 charging stations in parking lots though a government grant (Level 2, or L2, is typical of home chargers and takes 6–10 hours for a full charge). I remember telling my wife, Patty, that this was a terrible idea. And for some reason, I couldn't just let this news go. This would be a terrible waste of money and, by the way, some of those dollars were *my tax dollars*! I told her that I'd been thinking of a better way to charge these cars, so at that point I decided that I'd pursue the modular-exchange idea on my own.

Yes, the idea was still fuzzy in my head and very incomplete, but as anyone who's been involved in technology and innovation knows, you usually get a positive gut feeling when an idea has half a chance. I *did* have this feeling, so it was time to dive in headfirst. I was going to work on this invention completely on my own. I had no idea where it could all go, but the mountain was there to climb, and I was going to have at it. I will now get into the details of the basic problem I was up against, and the solution I came up with.

To start with, I'll describe the basic EV battery pack. EV batteries typically consist of around 100 large high-capacity cells in a series, giving a total voltage of around 400V (4V per cell). Sometimes there are one, two, or more of these series strings using medium-to-large-capacity cells. Each string is connected at the string ends in parallel to give the pack its needed capacity. In the Tesla architecture, each large-capacity cell is replaced by a large number of small-capacity cells in parallel. We can generalize that the basic battery pack architecture today is one or two series strings of large-capacity cells. This is what we were designing, and how I began to modularize the pack.

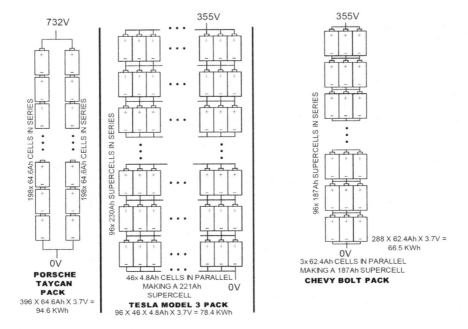

These are examples of battery-pack architectures. The Porsche Taycan and Chevy Bolt packs are typical of non-Tesla packs, which use larger format cells >60Ah for each cell. The Taycan ties the cells in a series-first approach and then connects two series strings in parallel. The Bolt ties the cells in parallel first, then builds a single-series string. Tesla uses small cells (4.8Ah) and ties many in a parallel-first approach (like the Bolt), then builds a single-series string.

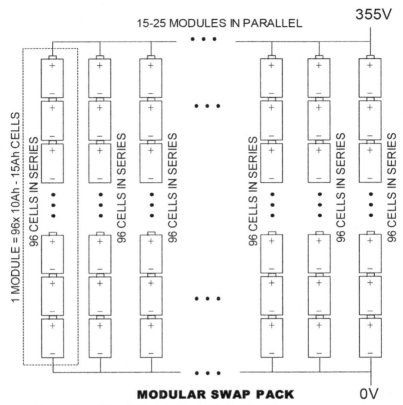

MODULAR SWAP PACK

EXAMPLE: 20 MODULES X 96 X 10Ah X 3.7V = 71 KWh

The modular swap approach is fundamentally different in its architecture. Like Tesla, it will use small-format cells (10–15Ah), then ties them all in a series-first string, much like the Taycan. Then there will be many (15–25) series strings attached in parallel. In modular swap, each series string will also have a "control element" that will modulate the DC current to maintain a balance of current in all the strings (not shown).

As I was saying, my first approach was patterned a lot like the conventional battery packs and used large-capacity cells in series. I would divide the series string into 15–20 smaller modules with 5–6 cells in each module that would be swappable. This had serious issues from the start. The current into *each* module is very high, and the connector would be a challenge. Also, if the modules were unbalanced or unequal, it would be difficult to equalize them. I wanted to allow for any kind of mixing of modules as far the charge level, chemistry, or usable capacity. Also, I wanted to allow for partial refueling where only a few modules could be exchanged and the pack utilized all the energy from each module.

The series module approach seemed doomed, but I kept hammering away at it for some months with no luck. Just as you really shouldn't use batteries of different ages and charge levels as you place them into your flashlight, you can't do that with series modules either. It proved difficult to work around this limitation.

Then came one of the big epiphanies in this whole process. Anyone who's invented something compelling is well aware of these "light-bulb" moments, and I happily and vividly remember this one. I was up in my mountain cabin staring at a cordless drill and its battery charger when it came to me: *That is a very simple connector,* I thought, *and it handles 15–20 amps easily. I need to limit my current, and I can do that by making 15–20 low-capacity battery strings, each at the full-pack voltage and tying them all together in parallel!*

Since the strings could be mismatched, I had to do something to effectively utilize all the energy from each module and also to prevent any current surges due to large mismatches in voltage as they're tied together. My task would now be to come up with a power-control circuit that could modulate the current level both in charge and discharge so that the current level in all modules would be controlled to be equal.

It all seems so obvious now, but as they say, hindsight is 20/20. I don't ever discount the fact that I'd already spent months brewing over other possible solutions. Pretty quickly (I would say in a matter of minutes to hours), the concept passed the initial smell test. Using lower-capacity cells wouldn't be an issue—that was what Tesla was already doing. The battery electronics would grow significantly, since there was the control portion and more series cells to monitor. My experience with electronics in the hard-drive world affirmed that circuitry cost also wouldn't be a showstopper.

CHAPTER 3

2011–2013 ARCHITECTURE IS DEFINED; PATENT IS FILED

I began an exhaustive study and design of the current control system in the module. This took months of modeling and simulation. I had very precise goals. I wanted to make sure that modules of any state of charge (SOC) and any state of health (SOH) could be mixed together. I wanted to be able to even mix different cell chemistries. Partial refueling was a necessity where only a small number of modules could be swapped and the SOC imbalance again would be compensated. The modules had to be safe: the voltage is quite high—near 400V—and when the module is outside the module compartment of the car, the terminal voltage should always be shut off.

I also realized early on that one other really nice benefit of using multiple modules in parallel like this is the high degree of redundancy in case of failure of one or two modules. If 20 modules are used, then losing 2 of those to failure would only reduce the max power available by 10%, enough to continue driving the car and to casually get into a swap station to replace them all with good modules.

The basic design was finished sometime in 2012. I used a basic pulse-width modulation (PWM) switching-circuit topology to build the current control circuit. I'd simulated the design many times under various conditions. One happy surprise I found was that when all the module currents were adjusted to be balanced at some total current—say, 100A—then the balance maintained perfectly for any other load current, like 50A or 200A. My simulations also showed that when mixing modules of different charge levels or capacities, the control system allowed the charge to distribute evenly among the modules in a way that allowed for full use of the energy from each module!

Maybe I should have expected all this, but *I didn't*, and I could see that this invention had some very compelling advantages. I had to get this invention disclosed for a provisional patent quickly, then pursue the utility patent.

My wife, Patty, and I spent quite a bit of time conducting our own patent search. By now I had issued around 30 patents in the hard-drive business and had a pretty good idea of how to go about the search and most of the patent writing. I was a bit surprised when we didn't find any clear cases of a similar invention in our search. It looked like we were in the clear. I would write the background but let my lawyer write the claims. I was very confident about this invention—confident enough to spend a significant bunch of my own money on a patent *without even having any idea where I was going with it*! I had no real plans on commercializing the idea myself, but I still felt that a patent attempt was an essential step.

Here is a copy of the patent abstract:

| United States Patent | 9,315,113 |
| Shrinkle | April 19, 2016 |

Electric vehicle battery systems with exchangeable parallel electric vehicle battery modules

Abstract

Systems and methods for electric vehicle battery systems with replaceable parallel electric vehicle battery modules are described herein. The electric vehicle battery system includes a plurality of electric vehicle battery modules connected in parallel. Each electric vehicle battery module includes a battery. Each electric vehicle battery module can also include a balancing circuit in electrical communication with a current path from the battery to an electric vehicle battery module output node. Each electric vehicle battery module also may have a current sensor in electromagnetic communication with the current path between the battery and the balancing circuit. The current sensor can be configured to sense a current level between the battery and the balancing circuit. The balancing circuit can be configured to balance the current level sensed by the current sensor of each electric vehicle battery module.

Inventors:	Shrinkle; Louis J. (Leucadia, CA)
Applicant:	Name City StateCountryType
	Ample, Inc.San Francisco CA US
Assignee:	Ample Inc. (San Francisco. CA)
Family ID:	50973885
Appl. No.:	14/138,683
Filed:	December 23, 2013

ABSTRACT OF PATENT

CHAPTER 4

2012-2014: TAKING THE TECHNOLOGY TO THE STREETS

In 2012, the company Better Place was executing its plan to make battery swapping a viable solution to vehicle electrification. But they were running into a host of issues. There was a growing movement against the whole idea of exchanging batteries as a refueling method. Of course, for myself this was all just noise. Unfortunately, this noise left a bit of a stigma when trying to discuss battery swap with anyone in the business. Better Place's failure was astronomical, affecting upward of $700–$800 million in venture funding. I started talking to automobile companies, as well as those making EV battery packs. The reception was less than warm and sometimes even hostile.

"BETTER PLACE RUNS OUT OF JUICE,
REPORTEDLY PLANS BANKRUPTCY"
—WIRED, MAY 24, 2013

I'd enlisted the help of a good friend of mine, Steve Yamamoto, who was working at a venture-capital firm at the time. Steve and I had both been physics students at UCSD from different eras, and Steve had a great handle on the interplay between the fine technical details of any technology and the business side of its implementation in the real world. We set out to license the modular swap technology, but we had a hard time convincing anyone of the virtues of modular battery swap—and we continued to hear this same old mantra: "Well, just look what happened to Better Place."

I started getting a bit discouraged—it didn't seem like anyone was really hearing what we were saying. They just didn't get it. I felt that there had to be a different way to do this, one that was more conducive to my life and travel style. My new thought was to start a "bottoms-up" approach. We were trying a "top-down" approach, going to the big auto companies, and it wasn't working at all. I believed that the idea needed studying and promotion, and that we could do that through a university partnership and project. This approach was kind of like changing things through a popular revolution as opposed to top-down policy changes. (I'm not sure that Steve was all that keen on this idea, but at my age and energy level, I felt that I could more easily hang with this tactic.)

I'd been involved with numerous industry/university partnerships before, so this felt natural and easy to me. One of my previous partnerships was at UCSD with Professor Raymond DeCallafon, who was an expert in servo and control systems engineering. Ray would be an excellent fit to work on this technology since the control methodology for the module was an area that needed some further study. Fortunately, Ray and his PhD student Xin Zhao were very interested right from the start. Our first plan was to apply for a government grant such as an Advanced Research Projects Agency–Energy (ARPA-E) or National Science Foundation grant to fund our work. Within months, we'd written a proposal and submitted it to ARPA-E with much confidence and very high hopes.

However, the reply to our grant proposal was anything but positive. Again, the reviewers were less than gracious, and I could see that these folks were just not listening to what we had to say. I then looked at the history of who *had* gotten these grants in the past and saw that these grants went out to the same companies and organizations all the time. I felt like we were just not in this loop, and I soon learned a lot about grants and lobbying, and the like.

Here we went again, and I began to ponder other ways of funding the research. There was this new method of funding I'd heard about called *crowdfunding*. Kickstarter, for example, had only been around for about two or three years at this point. This wasn't typically the way university research was funded, but why not? UCSD had a fledgling crowdfunding site for student projects of all sorts available for us to use, so I thought, *What the heck, this might work!*

Of course, we then had to convince people that we had a compelling project to donate to. We had plenty of basic boring technical research to go after with the funding, but I realized that we could involve senior-level engineering students and maybe come up with a very exciting and newsworthy goal that would really get the conversation resonating better with potential donors to the site.

How about a long drive in a car converted for electric drive and battery swap—like up and down the state of California? My son, Steven, a mechanical engineer, suggested that this was still too weak—we had to do something big. We had to make a drive across the country—*with the students*! I came around to the idea, but I knew that it was going to be a huge challenge. Then, Steven got quite involved in the project and was instrumental in helping us with the car conversion and battery builds. He also came up with a clever acronym/moniker for our project, which was M-BEAM (Modular Battery Exchange with Autonomous Modules).

It was pretty sweet working with my son on this project during those years. It will always be one of my fondest memories. He worked on it very diligently, and I knew that he had to truly believe in this technology at its core. That meant way more than all the harsh critiques and rejections we endured.

The managers of the UCSD crowdfunding site let us know that we should only expect to raise around $5,000 or so, which was pretty typical for most successful projects on that site. We needed way more than that, but at the time, anything would be helpful, so we went ahead with the crowdfunding site. We kind of shocked everyone when we eventually raised more than $47,000! We had to buy many battery cells, tool up numerous boxes, convert a 2002 VW Golf to electric drive, and buy chargers and rent generators and many other miscellaneous items to pull off this trip. The idea was to have two sets of modules and charge one set in a truck while driving with the other set. We would stop maybe 16–20 times across the country and quickly perform a 5-minute-or-less swap, then continue on. Students would be drivers, and part of the very physically demanding swap tag team. We finished converting the VW, and it ran beautifully. We ended up building around 40 battery modules, which was extremely tedious, as anyone who has built battery packs will understand.

The whole project lasted two to three years, and we had many accomplishments that I'm extremely proud of. There were three groups of senior students (14 students in all) over the course of the project who based their required senior projects on the M-BEAM technology. They were so awesome to work with. I remember each and every one of them with fondness, and I stay in touch with a few of them to this day. A few even landed jobs working in the EV and renewable-energy industries.

We also graduated two PhD students under Professor DeCallafon: Xin Zhao and Joe Jiang, whose PhD dissertations included very innovative work and research into battery swap. Also, multiple technical papers were written in academic journals. We drove the car quite a bit and began to plan our cross-country adventure. We needed to raise a bit more money, mainly for trip logistics—buying a few battery chargers, renting the generators, getting everyone home, and so on—and it was daunting! We struggled to continue the fundraising and had a few setbacks with some of the electrical and mechanical systems breaking down during testing, but we pressed forward the best we could.

I should reiterate here that the whole *point* of this cross-country trip was to get attention from the general public, and hopefully the business community, about the benefits of the modular form of battery swap. We worked with the UCSD Jacobs School of Engineering public relations department and became the subject of quite a number of magazine articles in publications such as *Top Gear* and *CleanTechnica*. We were also the featured stories on several local TV and radio news stations. Then, later in 2014, Ray got a call from someone in the Bay Area who'd read about our project somewhere and claimed to be working on the same sort of thing. A meeting that would change the course of our project and potentially the future of battery swap and EVs was set up in San Diego.

Khaled Hassounah had been involved in several successful high-profile, high-tech start-up companies and was quite active in bringing technology to developing countries in the world. He met with Ray and me at UCSD, and it was clear that his understanding of the benefits of battery swap and the inadequacy of charging stations was in line with ours. I liked that he wanted to take the bold step of commercializing the modular battery-swap technology. The goal of M-BEAM was not to commercialize the technology but to inspire and facilitate someone else in commercializing it, so this was a great opportunity to fulfill that plan. Building a business was neither in my skillset nor in my retirement plans.

The company, Ample, was formed by Khaled, and a license agreement was reached with me that included assigning the patent to Ample Inc. There was some negotiation to get to this point, but I recall that the deal was done after I asked Khaled why he wanted to do this. I remember his response was that first, it would be cool technology to work on, and he was looking forward to that challenge. But then he told me that the *planet* really needed this technology. I'm glad I asked that question, and the response could not have resonated with me more. I was all in, and my relationship with Khaled and Ample continues to be strong. We continued to work on the M-BEAM project, but I started shifting my efforts to help with Ample as much as I could to get modular swap off the ground.

Steven and I both worked with Ample early on to help them get their battery modules designed and the electronics in the battery management system working. Once Ample became better staffed, I took on more of an advisory role, which I continue doing today. Ample worked hard at the module development and the automated mechanical loading and unloading systems required for battery swap. After about seven years, they started making big waves in the EV industry and are now poised to expand their swap network, starting off in the fleet auto market.

"AMPLE IS LAUNCHING BATTERY SWAP FOR EVs, UBER DRIVERS IN THE BAY AREA FIRST CUSTOMERS"
—GREEN CITIZEN, AUG. 13, 2021

Yes, there are still many naysayers, and the conventional wisdom is still behind the charging-station paradigm. But besides Ample, there's a full disruption of that conventional wisdom happening in China, where its EV industry is starting to pose a severe threat to the auto industry that was controlled by the US, Europe, and Japan. It's pretty obvious that the government of China has become aware that the charging-station paradigm isn't adequate, certainly for their country, and that swap must be standardized and promoted. The Chinese leadership will absolutely make battery swap happen. But they're not using modular swap and are swapping the full pack, with all of its disadvantages.

"HOW NIO CHANGED MY MIND ON EV BATTERY SWAPPING"
—INSIDEEVs, JAN. 14, 2020

PART II

THE RATIONALE FOR A MODULAR BATTERY EXCHANGE INFRASTRUCTURE

In this part of the book, I'll lay out the numerous benefits of the modular battery-swap invention. I will also attempt to discuss the challenges of implementing the concept of modular swap. I use the word *attempt* because I won't deny the power of confirmation bias. During my tenure at two companies in Colorado, Storage Technology and Miniscribe Corporation, I had the privilege of working for Terry Johnson, who was the founder and president of Miniscribe. Terry was a brilliant leader, as he often elucidated his marching orders or his solutions through the use of parables involving such things as skiing, prairie dogs, airplanes, and so on. But Terry always said one thing that stuck with me, which is important here: "That's great, Lou, just try not to believe your own BS."

Those words served me well as an engineer and inventor, and I hope to apply them in this book. Just before the tumultuous year of 2011, Terry died in a solo plane crash in the Northwest Territories in Canada. We lost a powerhouse of an innovator and entrepreneur.

CHAPTER 5

FIRST BIG ADVANTAGE: BREAKING UP A PACK INTO SMALL MODULES

Before I get into the specific advantages of taking this large battery pack weighing 600 pounds or more and breaking it up into smaller modules weighing 30 pounds each or less, I want to take a look into the future and make what I think is a hugely compelling argument for pursuing a modular battery-swap approach and for everyone to continue to keep this approach in the front of their collective consciousness.

I have no doubt that in 10–20 years, the energy density of batteries will be much higher—I believe the density will be greater than 1,000 Wh/kg, which is 4X what the density is today. So batteries will weigh at least 4X less and will be more than 4X smaller in volume. Numbers much greater than 1,000 Wh/kg are consistent with the laws of physics and chemistry, which max out at 2,000–3,000 Wh/kg for lithium metal cells, or even 10,000Wh/kg for lithium air cells.

"INNOLITH CLAIMS IT'S ON A PATH TO 1000 Wh/Kg BATTERY ENERGY DENSITY" —INSIDEEVs, APR. 4, 2019

A 60 kWh (>200-mile range) battery eventually *will* weigh between 50 and 150 pounds, or possibly even less! It will be quite a struggle getting there, but we *will* get there. That is simply how technology works. I've been told that batteries are *not* disk drives, and this progression of size reduction will not happen. I say, "Wrong! They are." So one day we'll be able to refuel our cars by exchanging 15–20 modules, each one weighing 3–10 pounds and looking like a tissue box. This will happen in one or two minutes and could be done almost that fast even by hand! And keep in mind, the fast charger (Level 3 or DCFC—DC fast charger) will still take 20–30 minutes, and all of its disadvantages will remain. These Level 3 chargers operate from 50kW to 350kW power levels. I can't see how modular battery swap will fail to be the go-to method to refuel our cars, given this scenario.

To make it even more dramatic, let's say the whole battery is unrealistically very small and very light—80kWh (>300 miles range) and weighs only 5 pounds and is the size of a cordless drill battery. (No, this most likely won't ever happen, but this is just an exaggerated thought experiment to make a point.) To charge this in the car will still require that huge expensive DC fast charger and will take 20–30 minutes. But to swap the battery would take seconds even by hand, similar to your cordless drill. Dumb question: Which method of refueling would you prefer? I went through these scenarios in my head very early in this invention and came to the conclusion that this will eventually be the way that we refuel cars in the future.

Now we'll look at many of the other advantages that were either clear early on, or were pleasant surprises as they became evident much later in the development.

5A: SMALL MODULES ARE MUCH EASIER TO DEVELOP, MAKING VEHICLE ELECTRIFICATION MUCH MORE MANAGEABLE

During my tenure with ATL, I'd worked with a team here in the US building up large behemoth battery packs for EVs. I confess that we were a bit inexperienced with the challenges in prototyping these large packs. As I mentioned before, the large packs were just quite difficult to wrap one's hands around. Just the safety aspects of it all freaked everyone out. I laugh now when I think of the engineers and techs taking cover behind boxes and cabinets when we powered up the packs for the first time. But when I started thinking about a modular swap approach, it dawned on me that working with small modules could eliminate these fears. Prototyping could even be done at the module level on a desktop. Final integration into the full pack would happen incrementally and safely.

Since the modules would be standardized, a new design would really not be like starting with a clean slate. Each new module design would be leveraged heavily off the previous one. Development would be swift and very manageable. Engineering teams with a good deal of experience may find this argument a little weak since they have successfully built up packs. I'm sure an experienced team can manage prototype development in a very methodical way and avoid all the fear and explosions for the most part. But we don't always have the luxury of having such a team—and at any skill level, the modular approach will make the development much more manageable, which will certainly help accelerate development.

5B: MODULES CAN BE MONITORED—YOU'LL SEE EXACTLY WHAT YOU'LL GET

One aspect of the patent was that each module would be monitored continuously. The access to module data would be done wirelessly such that inventory at a swap site could be 100% monitored for such parameters as SOH, SOC, marginal cells, leakage, and any number of other parameters that are measurable and define the quality of the module. Some of this data can easily then be made available to potential customers in a way that they could view the quality of the modules at the exchange station using a phone app. Such a system would go a long way toward motivating improvements and quality control in the modules through transparency and competition.

5C: FREEING UP THE VEHICLE DESIGNERS

If the swap model is implemented by exchanging the very large pack, then it would seem that the car designers now have to work around a standard that would certainly hamstring them in the vehicle design. At least today, the pack is so big as to define the layout of the chassis and define the basic car without a lot of latitude. The use of small, swappable modules, however, can be placed differently with each model, and the number of modules can vary from one model to another to accommodate different vehicle sizes and costs. This will allow for the same standard module to support a wide variety in the fleet, from small sedans to large trucks. The designer does have the additional task of providing the means to automatically exchange the modules, and I will cover this later in the book.

5D: SOLVING THE BATTERY SUPPLY CHAIN

*"A LITHIUM SHORTAGE IS COMING, AND
AUTOMAKERS MIGHT BE UNPREPARED"*
—EMERGING TECH BREW, DEC. 13, 2021

If each auto manufacturer is left to solve their own battery-supply issues, we could potentially see some of these manufacturers left without enough supply to meet their needs. It's already happening that companies in China, and some like Tesla, are sequestering much of the minerals and raw materials needed to build batteries on the massive scale required. However, if modular battery swap was widely implemented, then this material "hoarding" would cease to be as much of a problem. Eventually there would be numerous manufacturers of these modules, and the manufacturers may or may not be automakers themselves. Then once these modules are in circulation, they would be available to any driver and any car.

The onus of gathering all these raw materials no longer falls on the automakers, which I believe is a very good thing. The Chinese will undoubtedly want to participate in the module business and will certainly want to build cars capable of modular swap, if the business is popular enough. The Chinese are already sequestering much of the necessary raw materials needed for batteries. This could be very helpful in not allowing the Chinese, or Tesla, to completely dominate the EV market.

5E: MODULE LIFE vs. COST vs. FUNCTIONALITY

Right now, with the "fixed large battery pack using the fast charge" paradigm, the requirements for the cell are fairly rigid. The cell must have a very high-powered density for fast charge and must last somewhere around 500 and 1,000 cycles so that degradation, especially when fast charging, does not occur early in the battery life. With modular battery swap, we don't necessarily have to ensure that the cells meet that cycle life as long as the costs are met and the module inventory quality is maintained, since any modules showing early degradation would be purged.

One of the biggest benefits I see that we will get by using battery swap is the change in design constraints for the cell designers and chemists. Cells could be designed to be much lighter and smaller since cycle life can possibly be relaxed, and the constraint to handle fast charge can be eliminated. Recycling and second-life uses (using degraded modules for stationary storage) could play a significant role in giving cell designers more latitude. With this added latitude in the cell design, it will also be possible to reduce the cell cost.

5F: VERY PAINFUL BATTERY-PACK RECALLS—NO MORE!

It pains me to be a witness to the current recall of the entire Chevy Bolt battery pack in all these cars. I feel really bad for all the engineers, line managers, and assembly workers at GM and LG Chem who are dealing with this mess. I own a Bolt, and I really like this car. This recall has not only damaged those companies, but sadly, will have a suppressive effect on anyone contemplating a move away from fossil-fuel vehicles. I've met in the past with engineers at LG and GM, and I know with absolute certainty that battery safety is priority 1, 2, and 3. There's no shortage of diligence in designing for fire prevention in the cells, the vehicle, and in the assembly process. As electric vehicles proliferate, it's a fact that more of these fires and battery-pack recalls will certainly occur.

> *"CHEVY BOLT BATTERY RECALL. HOW*
> *COULD THIS HAVE HAPPENED?"*
> — CAR AND DRIVER, SEPT. 13, 2021

But again, we encounter another serendipitous advantage of modular battery swap. Indeed, if there are safety issues with the battery modules, then the recall process will be substantially less painful—after all, removing the modules from a car happens all the time! A recycling methodology would hopefully already be established, and the suspect modules could be sent into that process to recover any of the costly raw materials in the module. Modular swap makes the removal of problem cells/modules from the field essentially painless, since it would simply be done at swap stations instead of at service centers.

CHAPTER 6

WHAT'S WRONG WITH CHARGING STATIONS?

6A: FAST CHARGE AND BATTERY DEGRADATION

The process of charging and discharging a battery cell is essentially a transfer of metal ions from one metallic plate to another, which accordingly will cause a flow of the free electrons in the opposite direction—thus, electric current. At first there's a large but finite number of metal (lithium) ions available for this transfer, but with each charge or discharge cycle, some of the ions get trapped and are no longer available for the transfer. This number goes up with higher current levels and with heat or cold. A fast charge will be more stressful to the chemistry than a slower charge, and will lessen the life of the battery, as measured by battery capacity degradation.

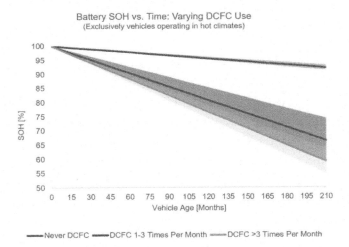

FIGURE 1 (source: *CleanTechnica*, Dec. 16, 2019)

Figure 1 shows how battery-pack degradation is affected by age, whether fast charge was used or not. SOH, or state of health, is the usable capacity of a battery cell compared to the specified capacity at purchase. The data shown was gathered in hotter regions of the country. It shows that capacity losses of 10–20% in a pack with the equivalent of more than 5–6 years of use are typical when fast charging is used regularly. This is not an insignificant amount of degradation. This could be as much as a loss of 60 miles of range in a battery pack with an initial range of 300 miles! This will have a stifling effect on EV adoption for some, as well as lowering resale values for EVs.

Who will want to buy an electric car with less driving range and possibly little life left on the prohibitively costly battery pack? And this degradation is only the typical amount. There will occasionally be some battery packs that will degrade even more, and all this degradation news will leave its bad mark on the EV business. This can obviously be avoided by employing battery swap, with its lower charge rates and continuous purging of marginal battery modules in the swap inventory.

6B: FAST-CHARGING PLACES: EXTREME DEMANDS ON VEHICLE AND CHARGING-STATION DESIGNS

As I step back and look at the present charging paradigm, it strikes me (as an engineer) that we're solving this problem by using a bigger and bigger sledgehammer. Tesla builds 100kW superchargers, then migrates to 250kW, and then others move toward 350kW fast chargers. The original concept has grown from this thing that looks a lot like a gas pump—probably no accident, as 350kW is a power-level equivalent to powering up 200 homes on average!

Yet it's nowhere near the delivery of energy that is accomplished at a gasoline pump. As a result, the fast Level 3 DC chargers are difficult to design and very expensive. They can tend to be unreliable, as many EV drivers are finding when trying to use them. The Level 3 charger requires a very heavy cable, or the cable must be cooled somehow, since the current level is high enough to significantly heat up the copper wire. The Level 3 currents must also be tolerated in the vehicle. The cables in the vehicle must also be larger than in a modular swap vehicle, meaning more copper and more cost. The battery itself is not 100% efficient during charging, and the losses will cause the battery cells to heat up. A very complex cooling system is required at *each* cell to keep the cells from overheating. This also adds cost and will affect the reliability of the vehicle.

EVs should have the advantage over internal combustion engine (ICE) vehicles in that they don't need any fluids to keep them running other than windshield-washer fluid. But with fast-charge, cooling is essential, and those vehicles will need this complex and unreliable fluid-based cooling system. Since DC fast charge is not needed with battery swap, it turns out that this cooling isn't necessary, and cable requirements are eased due to an often-overlooked fact about electric cars and fast charge, which is: Driving a car (or battery discharge) uses average power that's around 1/10 of fast-charge power. The peak power levels can be similar or even greater than fast-charge power levels, but peak power levels during driving are of short duration, while those power levels during fast charge are continuous. This means that the "heating" effects of discharge (driving) will follow the average power and will be 1/10 of fast-charge heating.

Anyone who's familiar with using Level 3 chargers is probably aware of another little annoying issue that eats into the realizable refueling range they'll see from the station: the Level 3 charger output power is *not* a constant through the whole refueling process, even at ideal temperatures.

"EV CHARGING IN COLD TEMPERATURES COULD POSE CHALLENGES FOR DRIVERS"
—PHYS.ORG, AUG. 1, 2018

The 250kW level, for example, will reduce as the charge level gets higher, to near 5–6kW levels during the last 20% of the charge session as a way of reducing the stress on the battery chemistry. And at cold and hot temperatures, the power level is reduced even more, such that the 250kW level may never occur. This means that refueling to 100% will take many more minutes, so most drivers just charge to 80% and drive right off, effectively giving them 20% less range. Another big advantage for battery swap: 100% range and charge is *always* available, with *no extra wait time*.

'Why did it take nine hours to go 130 miles in our new electric Porsche?'

—THE GUARDIAN, NOV. 28, 2020

"TESLA'S 250-kW SUPERCHARGER ONLY SAVED
US 2 MINUTES VS. A 150-kW CHARGER"
—CAR AND DRIVER, APR. 15, 2020

6C: WHY NOT JUST COUNT ON HOME AND WORKPLACE CHARGING?

We are nowhere near 100% adoption of EVs yet. When virtually everyone is driving an electric car, there will be quite a few changes in the daily rigors of personal transport. Home charging promises to be a convenient way to refuel your car, but access to such charging won't be available to everyone. Roughly, only about half the US population lives in a home where a Level 2 charger can be installed. Apartment dwellers, mobile-home dwellers, renters, and so on, will mostly be unable to charge at home. And households with multiple vehicles will find home charging to be a logistical nightmare. That being said, I find home charging to be a great convenience and useful for nearly 100% of my driving, as I'm lucky enough to have Level 2 electric vehicle supply equipment (EVSE) available. I have four solar panels in the backyard, and a battery system dedicated to only my vehicle charging, so the whole operation runs smoothly and is totally off the grid.

"THE HARSH REALITY OF OWNING AN
ELECTRIC CAR WITHOUT A HOUSE"
—CARBUZZ, AUG. 2, 2021

I also really like workplace charging; cars sit idle for many hours in workplace parking lots, and this would be at a time not concurrent with all the cars being charged at home at night. But there are also many challenges to workplace charging. It would be very unlikely that everyone could get access to a Level 2 charging station at some workplaces, since that would take the installation of way too many of these charging stations, and would require a significant power upgrade that would undoubtedly be too expensive for business or property owners. So for all those drivers who don't have workplace or home charging available, the only option is to visit a fast-charge station for their fill-ups, just as they would visit gas stations with an ICE car, but now with the longer wait time and living with increased battery degradation.

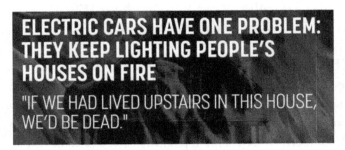

ELECTRIC CARS HAVE ONE PROBLEM:
THEY KEEP LIGHTING PEOPLE'S
HOUSES ON FIRE
"IF WE HAD LIVED UPSTAIRS IN THIS HOUSE,
WE'D BE DEAD."

—THE BYTE, DEC. 5, 2021

I do like charging at home, but the whole Chevy Bolt recall experience did give me some pause. The fire risk in all EVs is extremely low—the data shows that clearly—but for the unlucky ones, the results are extremely dramatic: the loss of your car and home, or worse. Since most battery fires occur while charging, most all of these would be avoided by using modular swap. I've heard from a number of people who would consider buying an EV, but they're holding off because of the fear of burning down their homes. *So why not give these people the option of using modular battery swap so they don't have to charge their cars at home?* Again, we need to find solutions for everyone so that we can get *everyone* driving electric cars.

6D: TO BE HONEST, WE NEED ALL EV REFUELING SOLUTIONS

I've pointed out some of the pitfalls of our current charging-station paradigm and how battery swap can in most cases do so much better. But I'm perfectly aware that these stations do work for many, especially home and workplace charging and the DC fast chargers for those on long trips with scheduled stops. I would propose that we don't throw that baby out with the bathwater, but instead, I envision a world where battery swap and charging stations will coexist, and drivers can use the method most appropriate to their lifestyles and driving needs. These needs will differ significantly from country to country and from urban to rural settings. Some will prefer DC Level 3 charging for trips, and many will charge only at home. But it's important to note here that vehicles that are outfitted for battery swap will be able to charge at home or the workplace with Level 2 chargers. They could also accept DC Level 3 fast charge, but likely at a reduced power level. The bottom line is this: whatever mix of refueling methods are used to get us to 100% full EV adoption is perfectly fine with me—that goal is just too important.

CHAPTER 7

IT'S SO NICE TO NOT OWN YOUR OWN BATTERY: THE SUBSCRIPTION MODEL

7A: SAY GOODBYE TO BATTERY DEGRADATION AND BATTERY FAILURES

Society is changing fast—it seems that younger generations are more comfortable with owning less and renting, and use sharing is a growing part of our culture. Cars are driven 5% of the time and sit idle 95% of the time—and much has been done to capitalize on the more effective use of goods, wealth, labor, and so on, so we have Uber and Lyft and VRBO, and millennials renting all their furniture..

Strangely enough, a persistent comment from people who really don't like battery swap is that they "baby" their cars and their batteries, and they don't want to be using those that could be tainted by others who don't do the same. I understand how this may have served them well in the world of fossil-fuel vehicles, especially with gas engines and their sensitivity to extreme usage. But the battery is inherently very measurable, unlike a gas engine, and as long as it meets its minimum requirements, it will perform nicely.

Batteries do degrade and they do fail, but remember, as long as the battery is easily swapped and the third-party owner keeps the inventory up to snuff, degradation or failure is truly nonexistent. The modules will be easily measured, and this data will be not only wirelessly available to drivers before the swap, but will be available to the modules' owners so they can effectively manage their inventories.

"TESLA OWNER BLOWS UP HIS MODEL S WITH
DYNAMITE OVER $22,000 BATTERY REPLACEMENT"
—ELECTREK, DEC. 23, 2021

Once an EV battery fails and it's out of warranty, an altogether new shock may be experienced by the car owner. The cost of the replacement will dwarf anything that car owners have historically been used to. A $10,000–$15,000 repair bill will be getting us into uncharted territory and will be another deterrent to full EV adoption. This can be completely avoided with modular swap and vehicle owners not owning the batteries. As George Foreman says, "Not my problem."

Right now, as I'm writing this book, I'm experiencing battery issues with my Chevy Bolt that drive home the point of "not owning your own battery" in a very personal and disquieting way. Up until recently, my Bolt battery typically showed around 250 miles of available range after a full charge. With the Bolt, I do notice that the number varies depending on my driving history and driving conditions, but typically I see 250 miles, plus or minus 10 miles. But now, for some reason, I'm consistently seeing around 230 miles. It could be degradation or some issue with the battery or battery electronics, but *fortunately, I do not own the battery*! I lease the car. If I *did* own the car, I must say I would not be feeling all that good about this, and certainly would begin the push to get GM to fix it.

Depending on the problem, the fix may not be all that certain or guaranteed—some degradation is allowed in the battery, and sometimes these problems are just dang hard to fix. I don't like the 10% loss I'm getting, but I'll probably live with it since I only have a year left on my lease, and the battery might get replaced anyway. This illustrates perfectly how modular swap—that is, not owning your battery, will make early battery degradation a completely different and totally tolerable issue for car owners.

7B: MODULAR SWAP WILL BE A PATH TO THE LOWEST VEHICLE COST IN HISTORY

At some point after I came up with the modular architecture, strictly to provide a more flexible way to refuel an electric car very quickly, it occurred to me that I was missing the main reason to pursue the swap approach. It was very exciting when I realized that the modular battery-swap subscription model meant that the entry-level cost of the car would be at least $10,000 less because, *Duh*, you're not buying the battery! This is a big deal. My goal was to get more people driving EVs, and this would really do it! I envisioned a world where the car would last for up to 1 million miles, the cost would be $15,000 or less for a very nice vehicle, and the monthly cost for fuel would be less than $100 for unlimited miles!

7C: SAVE THE GRID, SAVE THE PLANET—USE RENEWABLES

I did some thinking about how the subscription model would work—if it didn't pencil out, the whole idea would be dead, and I would be wasting my time. I'm no businessman, so my ideas will, of course, need to be vetted by those skilled in that art, but I can still apply my common sense and analytical skills to the problem.

From the start, I felt that the depleted modules that sit in the swap station should be recharged using the grid as the energy source. Those modules could very well be recharged at the swap station using grid energy, but that energy isn't cheap and may not be from a clean, renewable source! Solar could be employed at these charging stations, but that might increase the footprint in some cases so as to not be viable.

Instead, I propose that the modules be charged at a location where the energy is cheap and renewable, such as at a solar farm or wind farm, ostensibly located where there's more sun or wind and cheap land, but close enough to swap stations. The same would apply to geothermal or hydroelectric energy. Trucking could be done using electric trucks so that the whole process can be maintained with near-zero carbon output—powered by modules exactly the same as the ones they're carrying.

7D: AN ANALYSIS OF THE SUBSCRIPTION MODEL: IS THE MODULAR BATTERY-SWAP BUSINESS VIABLE?

In this section, I'll attempt to estimate the cost and viability of implementing a modular battery-swap business. To do so, I'll need to assess the total cost per month (TCPM) that would be incurred by the owner of the battery modules and the battery-swap stations. This may or may not be the same individual or entity for both, but we'll assume that the same person owns the modules and the swap station. It really makes no difference to the analysis. I will then get an estimate in similar fashion of the TCPM for the vehicle owner who would strictly be using DC fast chargers (DCFCs) for their refueling, and also the TCPM for the vehicle owner of a typical fossil fuel or ICE vehicle for comparison.

The typical gas station today will service around 100 cars per day. I'll assume that the swap station also services this same number and that the DCFC station also services this same amount. I will also assume that each customer refuels once per week and that the refueling is from empty to full. Since each customer refuels roughly four times each month, we can then deduce that each of the three different stations will service effectively approximately *700 customers each month* (which follows from the 100 cars serviced per day). We need to know this number of customers, since we'll assume that the swap customers pay on a monthly subscription basis. To meet these numbers, we assume that each station will have the appropriate number of refueling bays to accommodate 100 cars per day.

Another assumption that goes into the swap analysis is that to effectively service 700 customers, the module/station owner would need (700 x 20 modules/car) = 14,000 modules, plus some *extra inventory* to keep on hand and at the recharging facility. This number *will* be some fraction of the initial 14,000 specifically needed to go into the cars. There's absolutely no need to have as much extra inventory on hand as the total number of modules in the cars. The number I chose is 50% of the initial inventory, or 7,000 modules, which is very conservative. So, for 700 customers, the module/station owner must inventory 21,000 modules.

But the good news is that this *does not increase the total cost to the owner*, only the initial investment cost. This is because the modules will depreciate over a proportionally longer period of time since each module is simply being used (charge/discharge cycled) less than if the modules were fixed in the car. As a result, I will increase the depreciation time from 20 years to 30 years. It is also assumed that the swap station will require one truck trip each day. I've verified that the size of a semi-trailer should easily hold 2,000 battery modules, so the swap station can be emptied of depleted modules and stocked with 2,000 full modules with one truck trip.

MODULAR SWAP TCPM ESTIMATE

Assumptions:

1. Battery energy cost = $100 per kWh
2. Miles driven per driver = 1,000 per month
3. Module depreciation lifetime = 360 months
4. Wholesale electric rate = $.05 per kWh
5. Swap station = $200,000 (only 2–3 bays needed)
6. Swap-station depreciation lifetime = 240 months
7. Module trucking cost = $2 per mile
8. Trucking miles each day = 50
9. Swap-station labor = $4,000 per month
10. Swap-station land rental = $2,000 per month
11. Number of modules per car = 20
12. Module energy = 4kWh per module
13. Energy used per mile = (.25kWh)
14. Total number of modules per station = 21,000

Here are the estimates per mile of all the significant expenses incurred to the owner and operator of the modular battery-swap station: Note that all amounts are per month, and I added that cost divided up among the 700 subscribing customers.

1. Battery depreciation per month = (energy cost) x (# modules per station) x (module energy)/(depreciation lifetime) = ($100 per kWh x 21,000 x 4kWh)/(360 months) = $23,300 per month = $33.30 per month for each subscribing customer.

2. Total electricity cost per month = (miles driven per customer) x (# of customers) x (energy used per mile) x (wholesale electric rate) = (1,000 miles x 700 x .25kWh/mile x $.05) = $8,750 per month = $12.50 per month per subscriber

3. Trucking cost per month = (number of truck miles) x (trucking cost per mile) x 30 = (50 miles x $2 per mile x 30) = $3,000 per month = $4.28 per month per subscriber

4. Swap-station depreciation per month = (swap station cost)/(station depreciation lifetime) = ($200,000/240 months) = $833 per month = $1.19 per month per subscriber

5. Swap-station labor per month = $4,000 per month = $5.71 per month per subscriber

6. Land rental per month = $2,000 per month = $2.36 per month per subscriber

This gives a total monthly cost to the module/station owner of: $33.30 + $12.50 + $4.28 + $1.19 + $5.71 + $2.36 = *$59.34 per month for each subscriber*

Note: This is the module/station owner cost, and profit needs to be added to reflect the cost to the subscriber. For example, the profit margin selling gas today is pretty miserable: around 2%. Using a very generous 10% profit margin would allow a subscription to sell for around only *$66 per month*!

Next is the total cost estimate per month of a "customer" using a DCFC station. Here, there's no subscription, and the profits and costs to the station owner are built into the rate charged to the driver. This total cost will be the TCPM to the final customer (driver). Miles driven and energy used are considered the same as in the swap example.

DCFC TCPM ESTIMATE
Assumptions:

1. DCFC energy rate = $.35/kWh (recent DCFC average rate)
2. Battery energy cost = $100/kWh
3. Battery depreciation lifetime = 240 months
4. Battery pack energy = (20 modules x 4KWh/module) = 80kWh per car
5. Energy used per mile = .25kWh
6. Miles driven per customer = 1,000 per month

— Battery depreciation per month = (battery pack energy) x (battery energy cost) / (battery depreciation lifetime) = (80kWh x $100/kWh) / (240 months) = $33.30 per month for each customer

— DCFC electricity cost per month = (# miles driven per month) x (energy used per mile) x (DCFC energy rate) = (1,000 miles x .25kWh/mile x \$.35/kWh) = \$87.50 per month for each customer

So the total cost to each customer who strictly uses DCFC for all driving is: \$33.30 + \$87.50 = *\$120.80 per month.*

To estimate the TCPM for a fossil fuel or ICE car under similar conditions is much easier—assuming \$4/gal. gas and an average of 26 miles per gallon, we get:

(1,000 miles/month)/(26 miles/gal.) x (\$4/gal.)= *\$153.84 per month for the ICE driver.*

To summarize:

- Modular battery swap = \$59.34 per month (pre-profit)
- Modular battery swap = \$66 per month (post-profit assuming 10% margin)
- DCFC = \$120.80 per month (post-profit)
- ICE (fossil fuel refueling) = \$153.84 per month (post-profit)

One could certainly argue with my assumptions, but I did research them fairly thoroughly. The use of low-cost renewable energy from the energy supplier source for swap is a big reason for the cost difference. If anything, I believe that gasoline will continue to rise in price, and 100% DCFC will never be a good method to refuel EVs for those without home and workplace chargers, and its price will likely increase. Degradation will reduce the depreciation lifetime, and the cost of extra cooling and heavier cabling will add greatly to the \$100/kWh I was assuming for the DCFC battery energy cost. Besides, I'm pretty sure that anyone doing 100% DCFC will eventually get a little impatient with the time it does take and will cease calling it a "fast" charge. And as mentioned earlier, when using standardized modules, there will be further cost reductions for swap, with better recycling capabilities, more competition, and a much wider latitude in cell design due to the removal of DCFC constraints. The big takeaway here is that modular battery swap has the potential to be the lowest-cost method of refueling cars, and this is on top of the fact that the basic car entry cost will be at least \$10,000 lower, which will make a modular swap car the lowest in total cost of ownership (TCO).

We can estimate the TCO for the three cases over a 20-year lifetime if we assume that the ICE car and the DCFC car cost essentially the same—we'll use $35,000. For an equivalent modular swap car, we can assume that the cost is $25,000, or $10,000 less. Then, the total cost of ownership, notwithstanding repairs and other maintenance costs, is:

- ICE car = 240 months x $153.84 + $35,000 = $71,921.60
- DCFC car = 240 months x $120.80 + $35,000 = $63,992
- Modular swap car = 240 months x $66 + $25,000 = $40,840

What should also be obvious is that with modular battery swap, the resale of the vehicle should be significantly higher than a car with a fixed battery pack—even if the fixed pack shows little degradation. Most buyers will automatically assume reduced lifetime of the battery packs and bake that into the price.

7E: FLEET VEHICLES FIRST

Many who don't think that battery swap will become mainstream do believe it could have some application with vehicle fleets and trucking. Early on, it made some sense to me that working with fleets would be a more tangible way to introduce modular swap. This is how it's playing out with Ample, as they're working with Uber and others on their fleets. The benefits look unquestionable, with less vehicle downtime and lower refueling costs. I maintain, however, that what is good for fleet vehicles here will also be good for general transportation.

The retrofitting or design of vehicles capable of swap could be much more manageable at first with vehicle fleets. Swap stations would be run by the fleet owners and not for general consumption. Large-scale trucking should also benefit from a swap approach, since the charge times can be quite long, with battery packs in the hundreds of kWHs. Modules may be upscaled in size and power for large trucks, but that's not essential—only that enough modules be used in those large trucks to obtain the appropriate range. It may be best to use the *same* modules in a large truck that are used in the smallest car—all modules being the same equals lower cost by economies of scale. The semi-trucks carrying modules from the solar farm could, or should, use the same modules that they're delivering—one-stop shopping!

CHAPTER 8

IMPLEMENTING THE INVENTION: SWAP STATIONS AND VEHICLE DESIGN

8A: SWAP STATIONS CAN GO MOBILE! NO NEED FOR HIGH-POWER UTILITY SERVICE

Gasoline refueling stations have long been saddled with the difficulties of storing large quantities of gasoline underground. This is a very expensive component of the gas station, tethers the gas station to those underground tanks, and can sometimes present a toxic hazard due to leaks. Fast-charge stations are somewhat easier but will still be tethered to a very high-power utility transformer or even a substation. A modular swap station need not be attached to a fixed piece of real estate. It only needs to have sufficient volume to store enough modules to handle the necessary number of vehicles, along with providing enough throughput to be profitable in the area where the swap occurs.

The mobility of the modular swap station provides some new and interesting advantages for vehicle refueling. Consider the ability to relocate a modular swap station quickly to respond to either a seasonal or temporary change in traffic flow. An example would be providing for adequate numbers of modular swap stations along the routes to ski areas on busy weekends. Optimization of this sort will help manage inventory and allow station owners to relocate if the local traffic market dries up.

So far, the modular approach is the only one allowing for this mobility, as the full-pack swap stations tend to use belowground pits for the swap equipment. One criticism of a swap approach that relies on a lot of mechanical automation is that it would be difficult to operate in harsh environments. I would argue that the modular swap station is *better* suited for harsh environments.

DC Level 3 charging is very temperature – dependent. The charging rates will throttle to very slow at extreme temperatures, both hot and cold. This would, of course, not be an issue with modular swap. The modular swap station could be designed to be covered, as seen with the Ample station, which could be a nice benefit for customers in case of inclement weather. And solar panels could be integrated in. I wouldn't be surprised to see many Level 3 stations designed to also be covered to protect the cabling and the customer while charging anyway. It will be necessary to design the swap mechanics to be protected from a vehicle exposed to the elements, such as water, mud, and snow.

8B: WE WON'T NEED SO MANY MODULAR SWAP STATIONS—AND THAT'S A GOOD THING

There will be a fair amount of cost involved in a modular swap station. But *all* refueling stations are costly. A gasoline station with six pumps will cost somewhere around $500,000 to build—due to the expensive storage and pump hardware—and have a fairly large footprint. Fast-charge Level 3 stations are very expensive. These are extremely high-powered beasts running upward of 350kW, which would be the equivalent of the average power used by 200 homes!

The design of such beasts is quite a feat of engineering, using a large number of very expensive components operating with a high degree of parallelism. The cost of one 150kW station is around $200,000. The modular swap station cost will be dominated by the mechanical automation required to extract and install the battery modules. This could be near the cost of a DC Level 3 charging station, but it's important to realize that the throughput of a modular swap station will be 5–10 times faster than a DC Level 3 fast-charging station.

The faster throughput speed of the modular swap station along with the fact that a majority of drivers will be charging at home or the workplace for much of their driving will keep the number of modular swap stations needed to a minimum. The ability to go mobile and move with the market, as I previously discussed, will also keep the total of number of stations down. If we make some basic assumptions about the modular swap station's abilities and typical refueling, as it stands today, we can get a rough estimate of the total number of modular swap stations necessary.

ESTIMATE OF NUMBER OF SWAP
STATIONS NEEDED IN THE US

- Total number of cars on the road in the US = 290,000,000

We estimate that 50% will use home/workplace charging, so:

- The total number of cars in the US using modular battery swap = 145,000,000 (of course, this is likely a high estimate).
- The number of customers served by the typical swap station = 700. This swap station likely will have 2–3 bays.
- The total number of swap stations needed = 145,000,000 cars x 700 customers = approximately 200,000 swap stations. (For comparison, there are roughly 150,000 gas/diesel refueling stations in the US today.)
- Adding more bays to swap stations will reduce the total number of stations needed, and most gas stations have around a dozen bays by comparison.
- Many swap stations could likely be single bay and even mobile stations that would service fewer customers and perform fewer swaps per day.
- DCFC will continue to service some customers and also reduce the total number of swap stations required.

8C: A PROPOSAL FOR DESIGNING THE
SWAP-CAPABLE VEHICLE

Today, we find that the typical battery pack in a vehicle looks like a thin, rectangular slab that sits in the underbelly of a car situated between the wheels. The folks at Ample have built their swap system essentially around breaking up this slab into individual rectilinear modules and exchanging them from the underside by elevating the car. This may indeed be the best way to do so given that the vehicle designs today allow this as the only access. But given a clean slate, I would propose that a car have a series of rectilinear "tubes" that could either orient lengthwise in the underbelly or orient width-wise. The modules could then be exchanged though the front or rear of the car for the former orientation, or could be exchanged from either side for the latter orientation. This would not require the car to be elevated. The modules could be exchanged in the multiple tubes simultaneously, with total exchange times around 1–2 minutes.

In 1999, Dr. Art MacCarley, a professor of mechanical engineering at Cal Poly San Luis Obispo, wrote a very good paper detailing the battery-swap concept as a method of refueling electric vehicles. The paper was titled "A Review of Battery Exchange Technology for Refueling of Electric Vehicles," published by the Society of Automotive Engineers. It was an excellent work that obviously was way ahead of its time, and I recommend it highly. It's amazing to note that battery exchange was implemented first in the late 1800s—it is definitely *not* a new idea. In the paper, Dr. MacCarley describes the different methods for placing the battery in a car.

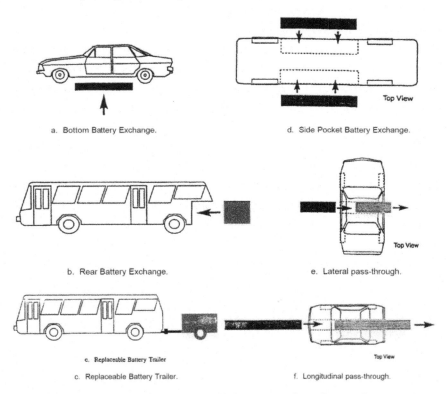

a. Bottom Battery Exchange.

d. Side Pocket Battery Exchange.

b. Rear Battery Exchange.

e. Lateral pass-through.

c. Replaceable Battery Trailer.

c. Replaceable Battery Trailer.

f. Longitudinal pass-through.

(Reprinted from "A Review of Battery Exchange Technology for Refueling of Electric Vehicles," 1999, Art MacCarley, Society of Automotive Engineers.)

Dr. MacCarley didn't anticipate many smaller modules being exchanged, but focused on exchanging one or two large packs. My preferred method of exchange would look more like either the side pocket, lateral pass-through, or longitudinal pass-through described by Dr. MacCarley. A pass-through approach isn't necessary when using multiple modules, but could be interesting as a way to minimize the exchange time. Ample chose to use the exchange from the bottom, I believe, as a more expedient way to work with present-day vehicle designs. Indeed, the side or longitudinal tube approach will require some new, innovative vehicle designs. But this shouldn't be an issue, since the required space on the car is available for this.

This tube approach would add some extra weight, but it should be very manageable, being well short of 100 pounds. The exchange tubes would also contribute in a very positive way to the vehicle structure and crash protection. I would propose that the tubes, for the most part, be made of structural steel, which would also provide some fire protection and isolation. The tubes could be cooled, heated, precooled, and preheated by diverting cabin-heater or air-conditioning air, which would allow the modules to be fast charged, possibly at somewhat of a reduced level. Also, vehicle operation in extreme environments would be considerably improved using this method of cooling and heating.

Also, the use of these tubes could potentially allow for a very simple way to "manually" exchange modules. This would be a benefit for do-it-yourselfers and for emergency roadside refueling by an AAA truck, for example. I realize that this self-reliance is kind of an old-school concept, but I grew up in a world where we rebuilt our own engines and changed our own oil. I must admit that I fancy a world where the car is still more open-sourced and accessible to the mechanically and electrically inclined—*and why not?* It would totally be possible for people in their own garages to build their own battery modules for their cars using discarded or open-sourced module hardware. Personally, I love the thought of this accessibility with modular swap even more than most of its other advantages. It would be a real plus for those of us who like to tinker.

The use of a set of "rectilinear tubes" situated in the underside of the car also presents us with a potential solution to help mitigate the damage from battery fires. The tube itself could be designed to "release" the modules to the ground below when the fire temperature reaches a specific level.

Once the modules are outside the vehicle, the fire can be contained easier. The vehicle could even be driven or pushed away from the burning modules on the ground. This would be accomplished by a selection of materials that would melt selectively and allow the modules to drop once the support structure goes away. Steel, for example, would not melt at lithium fire temperatures, but aluminum or plastic would.

8F: THERE ARE MANY OTHER AMAZING USES FOR THESE SWAPPABLE MODULES

"MORE POLLUTION THAN CARS? GAS-POWERED GARDENING EQUIPMENT POSES THE NEXT AIR QUALITY THREAT"
—KQED, FEB. 13, 2017

Besides propelling automobiles, energy itself defines many aspects of life in our modern society. This is becoming more and more the case—just think about how life comes to a halt when the power goes out. We should look at these swappable modules as not only a way to refuel an EV, but as a *small, portable, high-density source of energy*. All kinds of interesting scenarios can be dreamed of using these battery modules: There could be businesses that refurbish old spent modules. RV parks and RV enthusiasts could use these modules in clean/quiet generators and get refueled either at a swap station or with the camp host. They could be used as uninterruptible power sources for hospitals and other critical infrastructure. These "hot-swappable" modules could be easily used to build solar backup "power-wall" systems. The modules could fit on lawnmowers and other previously gas-powered equipment, further improving our air quality and lowering noise pollution. This is just a small sample of the numerous possible uses for these modules, given their size and accessibility.

"GAS POWERED GENERATORS ADD TO POLLUTION"
—BRUIN VOICE, NOV. 15, 2019

CHAPTER 9

MODULAR BATTERY SWAP AND
THE GRID

9A: A NEW WAY TO LOOK AT V2G

One of the really compelling ideas in the quest to use renewable energy and build a more sustainable world is the concept of moving to distributed energy as opposed to large, centralized energy sources. Basically, every rooftop solar system is a small generator in a microgrid. With enough of these small generators operating ubiquitously, we will rid ourselves of the large energy plants, such as coal-burning, natural gas–burning, or nuclear. Ideally, we can lose all the big smokestacks and won't need as many huge solar farms. But to succeed at this, we would need to also have nearly as many small energy storage nodes pretty much where the solar panels (or wind, etc.) would be deployed.

Green-energy experts would suggest that the electric car and its battery, along with any solar backup battery, would be connected to the grid in a two-way fashion so that some of the energy could be released to the grid at times when the sun isn't shining and the wind isn't blowing. Somehow, this would be accomplished with algorithms implemented on a "smart grid" in a way that works for both the homeowner and the grid itself. This is known as vehicle-to-grid (V2G).

It would seem that these algorithms could work but wouldn't exactly be foolproof. There *is* one significant downside for the homeowner/vehicle owner, however. The extra background discharging and charging of the vehicle battery that would be continually occurring in addition to the necessary charging and driving would hasten the degradation of the battery. But with battery swap, this degradation does not "belong" to the vehicle owner, but to the third-party lessor. It could then be accounted for by some allowance paid to the lessor by the utility company to allow the batteries to operate in a V2G mode. It's true that an allowance could also be paid to the owner of a fixed pack, but I would imagine that few people would be okay with increasing their degradation, even with some fair compensation.

"NEW STUDY FINDS V2G DISCHARGING
HARMFUL TO EV BATTERIES"
—CHARGEDEVS, MAY 31, 2017

By the same reasoning, V2G could be accomplished at the swap station itself, as quite a significant amount of power would be available there. Only a small portion of the modules would be required to send power to the grid, and an altogether different sort of algorithm would be required. Again, it would be the lessor who would own the degradation and would need to be compensated.

9B: CAN THE GRID POWER UP ALL THE LEVEL 2 AND LEVEL 3 CHARGING STATIONS?

"ELECTRIC CARS WILL CHALLENGE STATE POWER GRIDS"
— STATELINE (PEW) JAN. 9, 2020

"EV'S ARE ON A ROLL. HOW WILL US POWER GRIDS
MANAGE RISING DEMAND FOR CHARGING?"
—ELECTRIC POWER, APR. 24, 2020

"24 MILLION EVs IS THE LIMIT FOR CURRENT
US POWER GRID UNTIL 2028"
—INSIDEEVs, JULY 30, 2020

I thought it would be instructive to take a rough stab at figuring out how much grid energy would be required if the US went to 100% EV adoption, and refueling is only accomplished with Level 2 and Level 3 charging stations. I will do this using two different analyses. First, I will determine roughly how many Level 3 DC stations and how much total power is required to support the approximately 50% of those in this country who don't have access to a home or workplace Level 2 charger. Second, I will determine how much power is used by home and workplace Level 2 chargers and look at the loading on the grid.

9C: ENERGY REQUIREMENTS FOR DCFC SUPPORTING 50% OF VEHICLES ON THE ROAD IN THE US

Previously, we saw that there are 290,000,000 cars on the road in the US. If we assume that 50% of these do not have access to home and workplace charging, we then assume that they'll be using 100% DCFC. That makes 145,000,000 cars using DCFC for all their refueling.

To look at the power requirements, we will make a few assumptions:
1. Miles driven per car per month = 1,000
2. Energy used per mile = .25kWh

From these assumptions, we see that the energy required in the US per month is:

(1,000 miles/car) x (145,000,000) cars x (.25kWh/mile) = 36.25GWh
(36.25 billion or "giga" watt hours)

The total electric energy usage in the US is 300GWh/month, so this number represents roughly a 12% increase in demand. Other estimates I found put this higher, more like at 15–18%, probably because my mileage estimate is lower than theirs. This result does suggest that there may be some difficulties in some locations with grid capabilities, but by and large, the 12–18% change would happen slowly, and most agree it will be manageable.

9D: ENERGY REQUIREMENTS FOR HOME CHARGING

Next, I'll look at the impact of going to 100% EV adoption and how it will affect a local neighborhood. This will be instructive, and means a lot more than looking at the complicated picture of most homes and some workplaces installing and using Level 2 charging stations. The total energy required for home and workplace charging will be similar to 36.5GWh, since this energy is for the other 50% of the drivers, so again it represents another 12–18% increase in utility energy. Again, this increase will occur slowly, so it should be manageable. But there will be few disruptions to the grid, and they would tend to be localized. I'll look at a typical neighborhood here and get a sense of the level of possible disruption.

A typical utility transformer of 50KVA (50,000 volt amps, or essentially 50,000 watts) will typically provide power to 20 homes. This works out to around 2,500 watts per home and does seem to match well with the average daily usage in a typical home. During times of stress, when each home might be running air-conditioning at typically around 5–6kW, the load on the transformer can exceed its rating and run upward to 100kW, or twice the transformer's rating. The transformer will heat up, but these are very robust devices. Brown-outs and failures probably would not occur. With nearly all homes in that neighborhood (assume 100% home charging) charging their cars, typically in the evening hours, I expect that we'll see many overstressed transformers. Level 2 chargers run at between 6kW and 10kW and typically run through the night. Air-conditioners also can run through the night, so now with 20 homes, we could see:

(20 homes) x (8kW Level 2 chargers running + 5kW air-conditioning) = 260kW

Just the chargers alone draw 160kW from the transformer, but with air-conditioning also running, we could see upward of 260kW, which would certainly blow the transformer.

Of course, this will not all happen overnight, as I said, and the utility companies can upgrade transformers and make adjustments to the grid to accommodate these changes as they detect local grids running marginal. But all the changes made to accommodate Level 3 DCFC and Level 2 charging will ultimately need to be paid for. No doubt this will be accomplished by an increase in utility rates. I would propose that these grid stresses could be substantially reduced by implementing modular battery swap. The charging of the modules near renewable sources during off-peak hours and when energy is cheap and abundant will alleviate much of the grid stress and help keep utility rates down.

CHAPTER 10

LOOKS GOOD, BUT WHAT ARE THE PROBLEMS AND CHALLENGES WITH MODULAR BATTERY SWAP? (NO FREE LUNCH, RIGHT?)

10A: THE BIGGEST ROADBLOCK: RELUCTANCE OF AUTOMAKERS TO STANDARDIZE AND GIVE UP CONTROL OF THE BATTERY PACK

I'll be honest: since conceiving of the invention, it has always been on my mind that there would be a significant hurdle to convince the powers in Detroit and elsewhere to embrace the idea of building cars that would be part of a swap network using battery modules for which they had no part in either designing or manufacturing. This was reinforced in my early discussions with some of these automakers, and at times they weren't very gracious—okay, I get it. The automakers would lose one of their profit centers in the car: the battery pack. Redesigning the car to work with modular swap is a big deal, and they don't want to do it.

Now, these same automakers already make cars that use gasoline that they didn't manufacture, so is it so much of a stretch to view these swappable modules as very large gasoline "molecules" that don't end up in our air or lungs after being used to propel the car, but instead get recycled? I also get it that these automakers want to control the design and quality of the battery pack, since it really defines the car in the marketplace—both its performance and its value.

These are certainly reasons for automakers not to use standardized modules, but they're not necessarily *good* reasons. Flashlight or portable-radio manufacturers don't insist on building their own AA cells, and are accustomed to performing with cells that are trusted and can be purchased at any convenience store. Battery cells, again, are eminently very measurable, and the swappable modules will be way more transparent than even the batteries you buy for your flashlight at 7-Eleven—you'll know exactly what you're getting by just looking on your cell phone at the swap station!

This reluctance by automakers is actually one of the primary reasons why I've written this book—I want to educate people as much as I can. Why? Because I believe that good ideas will win the day, and the change can and will happen, not necessarily from the top, but from the groundswell of demand from the bottom, as long as the knowledge behind the change exists. I would encourage all of you reading this book to do your own research and form your own opinions about the claims I've made. I believe that the concepts surrounding modular swap and charging stations are very important parts of the whole EV revolution, and it is my hope that this book will be very instructive and will add to the conversation.

10B: WE'RE ALREADY TOO FAR ALONG WITH CHARGING-STATION PLANS

Ouch! I hear this one all the time. Right now we're in the throes of a political move where the Biden administration is proposing building around 500,000 charging stations at a cost to taxpayers of $7.5 billion. This pencils out to an average of $15,000 per charging station, which pretty much tells us that the majority of the stations will be Level 2.

"MOST EV CHARGING INFRASTRUCTURE IS
WASTED DUE TO LACK OF NEW THINKING"
— FORBES, JULY 28, 2019

But this is why I came up with modular swap in the first place! I couldn't see the utility in placing Level 2 stations in public places—the charging time is just too long to make using them an effective transaction. Unless these Level 2 stations are going into apartment complexes or into workplaces, this will end up being a gigantic waste, and I don't want the government to fail on this one.

In a perfect world, we'd be able to change course and not be the first ones to the finish line with the wrong answer. But all the work involved in building charging stations is happening now and will continue—and any battery-swap ideas are still a ways out from implementation, so we simply have what we have. But in reality, we should use any solution that works. Charging stations *do* work, especially in homes and workplaces, and should be used alongside modular battery-swap stations. If we're smart and build the right charging stations in the right places and then implement modular battery swap for the market that it will benefit, we can avoid a lot of waste and accelerate EV adoption.

10C: IMPLEMENTING MODULAR SWAP WON'T BE THAT EASY—IT'S A COMPLICATED MECHANICAL PROCESS

Of course, this is true—no innovation in the energy and automotive sector is easy to implement. There *will* be a need to come up with new designs for vehicles that will allow for automated machinery to extract and insert battery modules quickly, safely, and reliably. The modules would then have to be automatically placed and retrieved from a storage area within the swap station. Then there will be the machinery necessary to load and offload the modules from the truck at both the swap station and also the charging center, which would be located near the renewable-energy generator. These modules would have to be automatically moved and "plugged in" at the charging center, where they would be recharged at a safe, medium rate to 100% capacity. Yes, we will need *quite* an effort to make this happen.

But do we have the capabilities to design and build this infrastructure at a massive level? I say *absolutely* we do. For crying out loud, we drill the ocean bottom from these massive drilling rigs for the oil that's causing us so much grief. Just the infrastructure that supports these rigs, such as helicopter bases, tanker vessels and trucks, underwater drones, and on and on, will make the swap infrastructure that *will* replace the toxic oil industry we have now look trivial to implement! Essentially, we're just talking about building robotic machinery similar to what you might find at an Amazon fulfillment center, and this *will* one day replace all those massive ocean-drilling rigs, environmentally risky pipelines, and oil refineries that look and smell like something out of a movie about the apocalypse.

10D: SWAP WILL REQUIRE EXTRA MODULES TO BE BUILT, AND WE NOW HAVE A BATTERY SHORTAGE

There will indeed be a requirement that some amount of extra inventory of the battery modules would need to be built for modular swap to be effective. The exact amount isn't known, but it will not be a factor of 2X. The number could be as low as .15X or 15% added inventory, but will likely be in the 50% extra-inventory range. This certainly will mean that more of the raw materials needed for battery manufacturing would be required, and this could definitely put some stress on the supply chain. But I do foresee that this will be a short-term problem. Eventually this won't be an issue because the modules in service will last longer than fixed battery packs for a few reasons. One is that the modules could potentially last longer than the vehicle, meaning that the modules will continue to see service in newer vehicles after the old vehicles are scrapped. Second, since the modules sit for some fraction of their lives unused outside of the vehicle, there will be fewer discharge cycles over a given fixed time versus cells in a fixed pack. This will translate into a proportionally longer lifetime, and over the long haul, nearly the same number of battery cells will be used in modular swap as in using fixed battery packs.

I'm optimistic that we'll find the required amount of raw materials needed. Lithium, for example, could be mined from the ocean where there's almost an infinite supply, but right now this is too costly. Recycling could mitigate the need to continue mining the new mineral requirements by up to 95%. These issues are being worked on today, and I'm convinced that technology will solve these shortages eventually.

CONCLUSION

Right now in Taiwan and other Asian countries, there's a company called Gogoro that makes motorbikes with swappable battery modules. The concept is really taking off! People love it, and the riders get unlimited swaps (i.e., mileage) for just $25 (US) per month! Other motorbike manufacturers are signing up to the Gogoro module standard, and many see this as a total game changer for general transportation in the crowded cities of Asia. As of today, there are *more Gogoro swap stations in Taiwan than there are gas stations*!

Clearly, the modular-swap paradigm works for drivers and, while a car is not quite as simple as a motorbike, we totally have the capabilities to build up such a system for all general transportation. In this book, I've described an invention that will enable modular battery swap to be practical and robust in any car or truck. I hope that I've made a strong, compelling case for how this will help get everyone driving electric cars. I know there will always be some doubters—both cars and the auto industry are very personal, emotional subjects for many people. I welcome any debate on how we should pursue electrification. I'm sure that I haven't accounted for everything.

But most important, let's keep the discussion alive. There's a major disruption happening today in transportation and energy. Just about everyone, including myself, believes that this disruption is essential for the environment slowing down climate warming. But I want this mission to succeed and not stress our resources excessively. We need technology that will be effective and affordable. We can't rely on governments and legacy industries to make the right choices for us. Having an enlightened marketplace is always a move toward positive progress. I hope this book has taught you well, and will inspire some optimism with respect to the future of electric transportation.

ABOUT THE AUTHOR

Lou Shrinkle is a physicist, engineer, and inventor who has been granted more than 30 patents and has authored numerous technical papers. Currently, he is fully retired and enjoying life with his wife, grown children, and two grandsons in San Diego, California. A member of SanDiego350.org (an organization devoted to inspiring a movement to prevent the worst impacts of climate change), Lou advocates for clean transportation and pursues his own personal renewable energy projects. He loves playing tennis, bluegrass music, and going mountain biking.

Email: bluegrassrootsproj@gmail.com

Made in the USA
Monee, IL
03 March 2022

91878845R00039